FINDINGS

FINDINGS

An Illustrated Collection

FROM HARPER'S MAGAZINE

RAFIL KROLL-ZAIDI

ART BY GRAHAM ROUMIEU

With a Foreword by Patton Oswalt

TWELVE

New York • Boston

Twelve
Hachette Book Group
1290 Avenue of the Americas
New York, NY 10104

www.HachetteBookGroup.com

Printed in the United States of America

RRD-C

First Edition: November 2015
10 9 8 7 6 5 4 3 2 1

Twelve is an imprint of Grand Central Publishing.
The Twelve name and logo are trademarks of
Hachette Book Group, Inc.

The Hachette Speakers Bureau provides a wide range of authors for
speaking events. To find out more, go to
www.hachettespeakersbureau.com or call (866) 376-6591.

The publisher is not responsible for websites (or their content) that
are not owned by the publisher.

Library of Congress data has been applied for.

ISBN 978-1-4555-3049-6 (Hardcover ed.); ISBN 978-1-4555-3048-9
(Ebook ed.)

For Ming the clam (1499–2006)

Table of Contents

Foreword   *ix*

Findings   *1*

A Conversation with the Author   *77*

Acknowledgments   *105*

Citations   *107*

About the Contributors   *123*

Foreword

The first time I ever visited Pixar I saw The Table.

There, in a conference room, they've got this gigantic table. The kind you'd see in a movie about a sinister corporation, shot from a low angle in one of those slicing-up-the-world board meetings.

Atop the table is a jumble of whatever trinkets, gewgaws, knickknacks, and other objects the Pixar employees happen upon when traveling. Roadside keepsakes, souvenirs—anything that catches someone's eye, whether it's for the object's design supremacy or, more often, the accidental grace it carries despite its cheap, mass-produced nature. And not all the objects are manufactured—there are shells, and stones, and bits of wood and bone that, for whatever reason, made someone look twice, someone who after looking twice, in turn, made sure the object found its way to The Table.

The artists and designers occasionally drop by the room and pace The Table, letting the visual symphony wash over them, seeing if it jars loose a creative ember, especially when the grass of their inner, imaginative vistas has gone sere and yellow.

Many a signature detail in a Pixar film has its origin on The Table. Some of those are front and center on-screen—a sleek vehicle in *The Incredibles*, a doll-repair kit in *Toy Story 2*, a curve of reef in *Finding Nemo*. But some are happy to lurk in the background, to add emotional weight to a scene in a way you can't quite put your finger on but that stays with you long after the movie's over. Pay extra-close attention to the architecture of the dream city at the beginning of *Monsters, Inc.*, and then notice what's happening to it at the end. It's there—another story being told—but just at the edge of your vision, consciousness, and memory.

You hold in your hands *Harper's* version of Pixar's Table: at the back of every issue, and now collected for the first time, their Findings column. A jumble of facts, discoveries, and developments, micro and macro, in and about our world. Unspooling with seeming randomness, like a flow-of-consciousness Teletype from the center of the universe.

There are moments when you wander around The Table that subtle symmetries seem to tease you. Is that chunk of feldspar eerily similar, in shape and size, to that

Bakelite figurine? Is the font on that matchbook a first cousin to the veins on that palm frond? And on and on. Vanilla yogurt gives mice glossier coats and larger testicles. War is making Iraqi children shorter. Are these somewhere—no matter how far apart—on the same spectrum? We live in a universe that contains both. As well as us. So we're all connected. Yogurt, mice, and war.

This is how your mind will rewire itself as you dive into this book. And please don't just sit and read it from the beginning. Flip to a random page. Start reading in the middle. Let yourself move backward and forward, jump pages and paragraphs. Let yourself be a pond snail on crystal meth (they're in there) or a suddenly more charitable person at the top of an escalator (you'll see). Let this book be your Table. And send those embers out on the wind.

—*Patton Oswalt*

FINDINGS

Rude sales staff increase the desirability of luxury goods.

Americans who have just ridden an up escalator are twice as likely to donate to charity as those who have just ridden a down escalator.

Japanese scientists mapped the dopamine-based
reward system that encourages women, when looking
at themselves without makeup on, to apply makeup.

Danish doctors unveiled an
antidepressant helmet.

Kentucky is the saddest state.

Injury to the right parietal lobe correlates with a feeling of closeness to God.

The hearts of a Swedish church choir were found to beat in synchrony.

Scientists do not know why cranes sometimes dance alone.

Tylenol may reduce existential dread.

Particle physicists were optimistic about the possibility
of creating something out of nothing, because nothing
is actually something.

Swiss researchers induced in test subjects the sensation
of being surrounded by ghosts.

Children in a marshmallow study will eat the marshmallow if they believe researchers have abandoned them.

Children universally dislike clown wallpaper and find it "frightening" and "unknowable."

Girls are four times better than boys at growing up
with heroin-addict parents.

Rich children are better at filtering out
irrelevant stimuli.

Rich parents favor firstborn children
more than poor parents do.

Doctors found that some American children are prevented from playing outside because their parents dress them too fancily.

American children were eating more batteries.

A rise in anal sex among teenagers was noted by researchers at the Bradley Hasbro Children's Research Center.

A Croatian boy previously thought to be magnetic
was more recently thought simply to be very sticky.

Targeted social rejection activates the inflammatory
response of adolescent girls.

Babies as young as eight months enjoy seeing
bad puppets punished.

Bostonian teenagers who drink too much soda are likelier to carry guns.

War was making Iraqi children shorter.

The faces of Lego people have been growing angrier.

Evolutionary biologists proposed that men in the West had reached "peak beard."

Czech and German deer still do not cross
the Iron Curtain.

A Russian zoologist found that stray Muscovite dogs
have adjusted to post-Soviet urban life by commuting
from the suburbs on trains. The dogs prefer the front-
and rearmost cars and occasionally miss their stops
when they fall asleep.

An earless rabbit was born in Fukushima, Japan.

Scientists fear that the white black bears of western
Canada may disappear.

Columbus may have caused the Little Ice Age.

An Arizona cosmologist urged scientists to search for a "shadow biosphere" that may exist, undetected, alongside our own. Shadow life, it has been suggested, would be descended of a "second genesis" and would prove that life on Earth evolved twice over.

A Florida synthetic-biology lab announced the creation of a chemical compound capable of Darwinian evolution but said that the compound was not yet capable of living on its own. "It is not self-sustaining," explained the lab's head scientist. "You have to have a graduate student stand there and feed it."

Attentive fathers tend to have smaller testicles.

NFL quarterbacks play better if they are better looking.

Facial scars make men more appealing to women for short-term but not long-term relationships, with women preferring scars that suggest violence or trauma rather than acne or chicken pox.

Vanilla yogurt gives mice glossier coats
and larger testicles.

Scientists in New Guinea discovered a new genus of mice, which they described as "very, very beautiful."

A team of paleontologists suggested that dinosaurs developed wings to attract mates. "Maybe they ran around with their arms outstretched," said the lead researcher, "to show off how pretty their feathers were."

Researchers discovered how owls crane their necks. "Brain-imaging specialists like me," said a neuroradiologist, "have always been puzzled as to why rapid, twisting head movements did not leave thousands of owls lying dead on the forest floor from stroke."

A pox ravaged Japanese plums and one
Arizona hummingbird.

German police were disappointed in the performance
of Sherlock Holmes, a cadaver vulture, who confuses
animal and human remains and prefers walking
to flying; junior cadaver vultures Miss Marple and
Columbo, said the birds' trainer, "can't do anything
besides fight with each other."

It was determined that ancient Egyptians fed snails to
dead ibises.

Methuselina, the oldest ewe in the world, died in a fall from a cliff.

A Florida scientist definitively identified the G-spot of an eighty-three-year-old corpse.

In Scotland, most adults say it is acceptable for a man to marry his widow's sister.

Australian researchers were trying to solve the problem of humans outliving their eyes.

A third of male Londoners suffer from
penis blindness.

In western Iran, the growing popularity of *taqaandan*,
a pastime in which the top half of the erect penis
is wrenched sharply to one side and "popped," and
which has led to an epidemic of penile fractures,
was becoming a public-health concern. "The practice
of *taqaandan* is increasing," said urologist Javaad
Zargooshi, "and we don't know why."

It was revealed that British spies formerly used human semen as invisible ink; the practice fell from favor owing to the manifestation of a foul odor when fresh semen was not used and owing to mockery directed at the technology's inventor.

Conservativeness strongly correlates with a preference for name-brand mayonnaise.

Serving sizes in images of the Last Supper were
found to have grown by two-thirds over
the past millennium.

A newly translated Coptic text alleged Judas' kiss to
have been necessitated by Jesus' ability to shape-shift.

Teleporting a human into space at 30 GHz would take 4.85 quadrillion years.

Researchers turned a snail into a 7-milliwatt battery that can be recharged by feeding the snail.

The U.S. military reported progress in its cyborg-insect program and in building robots that can power themselves by eating the bodies of those they kill; the developers have promised that all "EATR" robots will be told not to eat people.

Ecologists in Wales feared an invasion by carnivorous blade-toothed eyeless ghost slugs.

The world's supercolonies of Argentine ants compose
a single ant empire that stretches across six continents.

Every year 100,000 sleeping Bangladeshis are bitten by snakes.

Scientists made graduate students provoke spitting cobras into attacking them.

Psychologists theorized that women's appreciation of true-crime literature may arise from a fear of crime and a wish to learn how to avoid it; it was further theorized that what women learn from true-crime books just makes them grow more afraid and buy more true-crime books.

Scientists proposed that male lions' skill at ambushing prey in dense vegetation was previously unknown because of scientists' fear of being ambushed by male lions in dense vegetation.

Test subjects experience fear when they are given a
third, prosthetic arm and researchers threaten that
arm with a knife.

Pond snails on crystal meth are better at remembering pokes from a sharp stick.

Korean researchers glued a dead bumblebee to a toothpick and made buzzing noises, scaring varied tits.

Five white tigers in a Chinese zoo had become fearful of the live chickens offered them as food.

Macaque couples were found to dislike the presence of
other monkeys of any rank during sex.

Bats were found to sing love songs.

Bees can remember human faces, but only if they are tricked into thinking that we are strange flowers.

Scientists concluded that it is particularly important to store box wine in a cool place.

A Conversation with
the Author

Tucked at the back of every issue of *Harper's Magazine*, Findings takes recent scientific discoveries and composes them as a three-paragraph column. Or maybe it's a three-paragraph collage essay. Or a prose poem. As much an act of creation as curation, Findings uses a kind of austere, lyric juxtaposition to turn what's essentially a compendium of facts into something full of wonder. From February 2013:

Happy adolescents become richer adults. Summer babies are less likely to grow up to be CEOs. Smart children are less likely in adulthood to report chronic widespread pain. Autistic children take longer to learn to be afraid of new things. Many Swedish children who self-harm don't really mean it. Lying increases the temperature of the nose. A wandering mind shortens one's telomeres. Fetuses yawn.

As a devoted Findings reader, I always feel simultaneously more sure and less sure about the world we live in by the end of each column. And as a writer, I'm continually impressed with the way straight expository writing can be put to such creative use, evoking such a wild array of emotions within a single paragraph, sometimes a single sentence. Findings runs without a byline, but there is a single person behind every month's column from start to finish, so I sought out that person: Rafil Kroll-Zaidi, a *Harper's* contributing editor and at the time its managing editor, who has written Findings from December 2007 through the present. We spoke in August 2012 in a café on the Bowery and later over Skype.

—*Dave Madden*

Findings is a fairly new feature of a very old magazine. What was the thinking behind its genesis?

Findings was an invention of Roger Hodge, who was the editor of *Harper's* from April 2006 until February 2010, and who then edited the *Oxford American*. *Harper's* is very far from having anything like a house style. There are certain forms and certain postures and certain concerns that crop up again and again, but the way that the comma is used or a piece of reportage is structured from one writer to the next may vary enormously. At the time Roger introduced Findings there were three areas of the magazine that helped to form the house voice. There was the Web-only Weekly Review newsletter, which is bylined, but it's always written in the same deadpan style. It's always three paragraphs, the same way Findings is. There was the Index, which is not bylined and also very deadpan. It has the same not very veiled irony. And there was the Readings section, which is more curatorial, because it's all excerpts and the editors just write the introductions. What Roger did with Findings was create a further expansion of this *Harper's* voice, where you use discovery and juxtaposition, and where the editorializing tends to consist in the ordering of things and in the choice of what to talk about. But within the sentence itself everything is played quite straight.

Those old columns felt more newsy. There'd be more direct quo-tation from the study itself, say, and not as much juxtaposition as you do now. Was that a conscious change you made when you took it on?

When you have an inherited form that's not bylined, you have an institutional responsibility, and you don't mess around with it that much. It would be interesting to go back and see how a Roger Hodge column from 2006 compares point-by-point to one now. You'd notice small things. The changes in syn-tax and punctuation. Different kinds of cheap jokes. Different sets of recurring concerns or interests. There were a lot of...not missed but deliberately untaken opportunities for funniness—especially in the lesser degree of obscurity and of convolutedness. But the basic, kinetic, punch line–seeking struc-ture of the sentences, to take an example, has been extremely consistent. I'm vaguely recalling not an actual line here but a piece of information from an old column, and I remember it as: "A cat gave birth to a dog in Brazil." Now, what's interesting is that a cat gave birth to a dog. "In Brazil" is just dead lan-guage sitting at the end of the sentence. You don't care about it by the time it happens. So if that were my line I'd go: "In Brazil, a cat gave birth to a dog," so you've got the kicker at the end. Or...? Better...? Can you guess?

Um.

"A Brazilian cat gave birth to a dog."

Okay.

Because it's worth getting Brazil in, but it's not worth having the end of the sentence muffle itself and trail off.

Right, that I get.

At the beginning of the sentence it's just a necessary adverbial bit of information. But when you shift it to that adjectival mode, then something goes slightly queer about it. Because there are some species where the common name will have a name or place-name attached to it. Geoffroy's cat or a Siamese cat. A Russian wolfhound, a Hungarian sheepdog. So for one thing it starts to generate a little bit of tension because it's being misused in that way. And then also, if you put that before anything that has to do with the female anatomy, then you have *Brazilian*, obstetrics, and "pussy" all falling in the same sentence. So there's a lot that you can do with a little. The greater confusion of that sentence is that this should not be happening. A cat should not be giving birth to a dog. But you can do that further defamiliarization by saying, "Does this cat have a nationality? Is the pertinent fact here that the cat was Brazilian?" It's just the idea of being in Brazil is much deader than being…

Of Brazil.

Yeah. Being of Brazil is very different than being in Brazil, and being of Brazil in that situation seems to open up a much larger set of possibilities for innuendo, for anthropomorphism, for taxonomic confusion. There's a lot more play at play when you do that particular line that way. And it turns out, now that I'm looking up that original column, that the line we wound up with here is *exactly* the line as Roger wrote it.

The older columns were a lot more categorical. There'd be a paragraph about astrophysicists and space stuff. And then one about animals. Whereas now there seems to be a lot more play in moving among topics within paragraphs than there was.

Yeah, part of that is the pleasure of making different kinds of turns. So you start with something that's about sperm quality and that moves to something that's about fetal development and then you move to something that's about premature birth. (I think I may actually be describing an arc that I've used.) Then you move to something that's about childhood development, then you move to something that's about how child development will diverge depending on the race of the child or on the economic circumstances of the child, then you may move to something about adults of a particular demographic—African American men are more likely to suffer from sleep paralysis, or rich Americans are more likely to sue their doctors—and from there you'll have further steps, but you can follow a very particular progression where each item's relationship to the one that preceded it is clear. Or you can have something where there's a transition that's just incredibly cheap. Like the Spitzer Space Telescope to prostitution. And then you can put astrophysics before that pairing and you can have sex and relationships after. I think the way that I actually wrote that one was probably Spitzer–hookers–blow, but I could've split a paragraph neatly into two sets of topics with one very cheap trick. Then of course one of the pleasures of the form is the non sequitur, which has a very particular flavor in Findings.

How would you characterize that flavor?
Often there is this expectation that a paragraph will end on something that is a non sequitur, that diverges substantially from the general tone that precedes it. It's a very classical use of certain devices. You can have a bathetic effect with the

non sequitur. So you move from cosmology into scatology, for example. Or you flip it the other way. You can have the most somatic, gross-out sort of sticky human concerns and then the final thing would shift completely out of that realm and be one short sentence about a radio signal from a star 27 light-years distant.

Is that just fun to do, or do you think there's an overall effect Findings *is going for that that move helps achieve?*
The self-consciousness of the form induces in the reader ideally a sort of consciousness of his or her own values. The non sequitur pairing I described is a bit of a memento mori, which is often an effect Findings goes for.

What are the differences as you see them between a Harper's Finding, *a scientific finding, and a fact?*
A scientific finding is inherently useful whether or not it involves a definitive result, because it's part of the grand ongoing project of acquisition and dissemination of knowledge as conducted by humans in the universe. A Findings finding is interesting if it's interesting, if it's an encapsulated simplified version of some discovery about or event in the phenomenal

world that withstands multiple tests of human boredom. And by human boredom I mean the boredom of a single human, which is me.

But when it comes to the column's disseminating facts, what's your responsibility? What kind of fact-checking do you do?

We tend to take at face value the press releases whatever lab or university has issued. Sometimes—either for reasons of lack of clarity about process and data or just sloppy writing or excessive brevity—an article or press release won't be sufficient and I'll read the original study, if it's out, or contact the individual scientists. I think the only time we've not been able to do that satisfactorily was with a study about sippy-cup injuries to toddlers. It was a very comprehensive study. It said that 70 percent of such injuries involve the mouth, 20 percent involve the face, head, or neck, like you would expect. And it said less than 1 percent of injuries involve the leg, the groin, and the kidney area—something like that. Obviously what I want to know is what percentage of injuries are to the groin. At first the study authors were cooperative. We didn't even ask specifically for the groin data, we just said, "Of the things where the percentages were so small that you didn't list them separately, you must have specific numbers? Can we just have those actual numbers?" And they kept waffling and going back and forth, and after three days they came back and said, "We don't want the data to be misused." Which really pissed me off because all the other data they had made available.

And how would you misuse them? Incite more sippy-cup violence?

Right. And then sometimes what you want to say requires further calculation. So if 3 percent of U.S. automobile owners suspect their cars might be haunted, then you obviously want the total number of cars that are suspected of being haunted. Then you have the problem of individual persons who own multiple cars or multiple persons who jointly own individual cars. If you can fiddle your way through that, and you can say maybe 210 million automobile owners in the United States, you end up with roughly 6 million haunted cars. For this hypo-thetical study they probably would not have inter-viewed enough people for us to apply that to the population as a whole with that kind of confidence, but that's one of the liberties that would be taken.

I want to push this, what you said about obviously you want the number of haunted cars. Why, exactly? What end purpose does that fact better serve than the percentage of cars?

Would you tend to agree with me?

I think I agree with you. But I can't figure out why.

With that statistic in particular, what you're put in mind of with the percentage of car owners is that there's a number of people who are a little batty and superstitious about certain everyday objects. Whereas if you can get the number of cars that are suspected of being haunted, then you know you're driving down the highway with ghost cars. One thing is a scientific finding, the other thing is a grave spiritual concern.

You had said something about the ways in which the style of Findings is kind of an imitation and celebration of scientific study.

I didn't say of scientific study. I said of scientific reporting, which is usually bad.

Bad in what way?

Well, the bad habit that Findings is directly in conversation with is the news-driven—now more specifically, page view–driven—definitive, declarative, pithy statement of research results. The column gently parodies those headline-only articles by imitating them, frequently with a tweak: in the many staccato sentences that constitute a column there's alternately a lot of passive voice and a lot of actors doing the finding. *Someone* is always responsible for the experiment or the observation or the reporting, and all those processes involve design and selection and narrowing, which I guess I then do as well. In Findings the immutable statement of isolated fact, especially when it relates to experimental study, is often a counterpoint, punctuation. One of several styles. There's actually a great deal of caution and weaseling for a form whose presentation is aggressively cavalier!

The headline-ese problem is well known, but it still interests me, particularly the subproblem of the media's love for confectionary social-science studies that have very small sample sizes. I myself love these. So ten female college students pick from an array the photo of the man whom they find least threatening, and then suddenly it's like, "Ovulation makes straight women less afraid of masculine men!" Well, that's a combination of the actual findings plus two or three less supported extrapolations plus an offhand statement the lead author of the study made to a reporter or the university's press office. Or take experiments in behavioral economics. Those usually involve extraordinarily abstract setups and operate via psychological metaphors that are at multiple removes from reality. This red poker chip is "market price" for that banana, a $5 bill is "being wealthy." So what such a study produces is this attenuated, 32-bit Amiga simulation of what human beings are like, and from that come some interesting implications, but when the story appears, it's talking entirely about the implications of the implications. This isn't to say there aren't persistent problems in long-form writing about science for educated audiences—there are, and that's territory I know from the inside—but that's not what Findings is sort-of-parodying.

Was that parody part of the spirit of it in the beginning?
I think it was.

Because in an early column of yours I think there's an end-of-paragraph non sequitur that reads, "Men are both smarter and stupider than women."
That would have been December 2007, the first column I wrote.

And I think of that as a classic Findings line, how it doesn't say anything about anything, and yet it says a lot about what we believe.
Sometimes what'll happen with the press release about a study is that there's almost no way to rephrase it. So everyone who reports on it says the same thing, including me. But very often you get a sense of when

very boring studies are going to contain something interesting. I can run through a list. There's finding counterintuitive to fact. Confirmation of the obvious—there are ways to be incredibly reductive though accurate; there might have even been a study about this, which found that black people tended to agree that George W. Bush doesn't care about black people, something like that, where it would've been a nuanced, carefully worded, rigorously carried out study by an independent polling center or by a major research university, but, you know, five months after Katrina they called up a bunch of people and that is actually the essence of what they found. And then there's human–animal conflict. Pathetic fallacy is another. That's similar to anthropomorphism. I used to have a lot more sex in the columns.

I guess there hasn't been as much lately.
No. There used to always be an entire paragraph about sex, and I think this has changed because of my switching from Google Alerts to RSS feeds.

Really?
I was also maybe just getting sick of it, but I think that was the thing. It used to be that I'd get Google Alerts for *sex study* and *behavioral study*, and I would get separate alerts for every kind of inherently funny

animal. Inherently funny animals are still a favorite subject. They can be funny either because of the name or because of the animal itself. Like a mongoose by any other name would not be so funny, but a platypus by any other name would still be funny. Platypuses are almost cheating, because everyone knows platypuses are funny.

What are some other funny animals?

Macaques, because they're monkeys and many have red butts. I've been actually harassed by macaques on numerous occasions.

Oh, wow.

When I was a child, they'd come into the house and steal bananas, steal my diapers, or things like that. Bears are inherently funny.

Are they?

I think they retain some of their totemic fascination. There's no way to reconstruct the original Indo-European root for *bear* from any Germanic languages because to speak the name of the bear was apparently taboo. Words for bears in Indo-European languages are euphemisms, though the euphemisms themselves may go back to an Indo-European word, for "the honey-eater," "the brown one"—as with *bruin*, that sort of thing. Even the supposedly original word, **rkthos-*, which persisted in languages like Greek and Latin, may be a euphemism: for "the destroyer." Another one, not funny but innately compelling, is bees. Bees had privileged status in classical antiquity. They stood for an ideal of civilized order. And they're important for early Christianity because they practice parthenogenesis, or maybe it's that parthenogenesis is observed among them.

Other common subject matter is just a bias that results from my choices of reading material: for example, so many silly and twee things tend to happen in the U.K. because there are so many funny place-names and there's this kind of Tolkien-esque quaintness about the proximity of certain cute animals to civilization. There's not this same stark progression from city to suburbs to countryside there. So you have all this stuff involving hedgehogs and badgers in places that have adorable English or Scottish or Welsh or (Northern) Irish sounds.

Lots of y's or something?

The Welsh ones are funny because they're unpronounceable. The English ones are funny because, like, "Hug Me 'Pon-the-Tyne" would be the name of a town. In these places the human-animal interactions are instances of benign concern or symbiosis, whereas in America animal stories are so much more often about a rattlesnake dropping out of an acoustic-tile ceiling. And in America this happens in part because of the rapid expansion of exurbs into totally inappropriate places. Los Angeles. Florida. I mean, *The Orchid Thief* is like a catalogue of how inappropriate it is for modern humans to live in most of Florida.

If Findings were an animal, what kind of animal would it be?
  Uh…A rara avis?

Nice try.
  I mean, is Findings an animal that's capable of tool use? Is it a raven? Is it a dolphin? I would say that Findings is a dolphin. It's an animal of tremendous intelligence and versatility that's constrained within the inarticulate joy and rage of its water-lockedness and its lack of opposable digits.

That's pretty good.
  I wish it weren't an animal that was so '90s, but, there you go.

I guess this is a place to ask about process. Do you keep files?
  All of it's done in about a day.

The whole column?

It happens pretty fast. It depends on how interesting science is that particular month. Sometimes science is pretty boring. I let my science-news feeds build up over a month. Then I go through and star anything that might be interesting—whether that interest is something that's obvious from the headline, or whether it's an article that appears boring but that there could be something ridiculous buried in it or in the study itself. Usually I look at the headline or the summary text for a little over 1,500 articles, and I wind up starring maybe a couple hundred. Then I go through and read the articles and skim the studies, and I start writing the lines in a rough form. Part of the test of whether I keep something is whether the line works at that stage. Or whether they come together. If I start noticing eight different things about penguins, I'll type the penguin lines near each other. I'll get a little penguin block. Then I'll go through and start cutting and organizing. Because it's such an intensive all-at-once process that very quickly leads to boredom, whether I keep something depends solely on whether looking back at it a few hours later or the next morning it still seems interesting. That process of culling is also a process of matching. So if a line seems to pair well with another line, I'll move that over. Maybe eight or ten blocks will start to float together as it gets cut down, and by the time I have those blocks it usually gets down close to the word count, which is a little over 500 words.

So all that's a day.

It's about 16 hours of work, depending on the state of science. And yeah and then it goes to the fact-checker. But again it's basically a clip job. And citation in the magazine would undermine the column's confusing charm as a prose poem.

How much of the character of Findings is your character as its author, and how much belongs to the form itself, as you inherited it?

To the extent that it's about me, it's about what bores me. Or what I do to stay interested. I've gotten into the habit of trying to provoke people with the way that Findings lines are phrased. For example, many young Israelis travel to India after they complete their military service. They go backpacking and get stoned. So the Israeli government issued a travel warning that said there'd been a number of rabies outbreaks in places frequented by Israelis in India. The line that I wrote in Findings was, "The Israeli government feared lest its citizens become rabid." Which is the most grotesquely editorializing thing

I could do. Surely someone will write in about that, I thought, but nobody did. The only thing to date that got anything approaching a really notable number of complaints was when I said that vegetarianism may cause brain shrinkage.

What? Brain shrinkage?

Yeah. And the vegetarians lost their minds—because of course their brains were shrunken. But really, they lost their minds because the study was actually about how those who don't eat various sources of animal protein are at high risk for B-12 deficiencies and therefore brain shrinkage. But it was phrased in a very reductive way and those people just went bananas.

So is Findings mostly just a bunch of jokes?

It's equal parts to inform and to amuse, and these kinds of facts don't have to be forced very hard to do both. They naturally lend themselves to both. To the chucklesome and to the sublime. So that seems to be something that the form exploited almost from its inception. It's actually hard to say whether that's because there's something deeply funny about these kinds of discoveries, or just that the capacity of them to be funny has been underused. I mean, you take a realm like politics. The humorous potential of the news there has been exploited tremendously. What I've been exploiting more and more is the limitless opacity of certain scientific findings. (Many of them don't meet the standard that grant seekers by and large are terrified about, which is a kind of amorphous vested-interest standard: "I'm a hematologist. Can I get money from the military to study an obscure form of childhood anemia because it's possible that this would have some application to soldiers on the battlefield and how we might be able to promote clotting in wounds?") But there are some things that it might be possible to explain, but it's actually a lot less fun to explain.

Almost like an art-for-art's-sake aesthetic but for scientific study and research?

When you have a line in Findings like "U.S. Department of Energy researchers broke Kasha's Rule," I'm not going to explain what Kasha's Rule is. When you say—when I say—that scientists hope to harness the Casimir Force, some people who took some advanced physics class may remember what the Casimir Force is, but I'm also not going to explain that.

Well, what's your aim in not explaining it? Is it to maintain a kind of strangeness?

It maintains a kind of strangeness and also it's about the extent to which science can observe things or offer explanations. When you have "The Bruce Effect was documented among wild gelada monkeys," I don't bother with who Bruce, the woman, was or what her effect is, but the Bruce Effect might just be something where, you know, grooming tends to be proffered by subordinate males to dominant males rather than demanded of subordinate males by dominant males. I don't know. Part of this is the process of becoming bored by explaining. It reminds the reader that—actually, a very talented and technically literate writer I've worked with, Hamilton Morris, said it well: Much scientific research is fascinating precisely because it is *not* intended to appeal to a general audience.

I think the people who read this column and understand and appreciate it are over the idea that science takes place for science's sake, to the extent that this idea both is and isn't true. Findings celebrates the idea that modern science is a tremendously powerful and productive and beneficial and motivating and clarifying force, but the idea that everything that goes on is part of this heroic, conclusive, triumphal narrative is also silly. You know, the universe defies and denies and startles and confounds us just as our own bodies defy and deny and startle and confound us. Findings' being funny is partly a corrective to that particular form of triumphalist narrative.

One of the things I like about scientific findings is that they seem to be presented as the end of a story. But in your column, findings become simultaneously the end and the beginning of the story. An example would be the Brazilian cat, or that goose in Arkansas named 50 Cent who survived being shot? Again these things have a resolved feel, and yet all they do is suggest suggest suggest, and you can't help but think, Wait, so what happened next?

Yes. I'll do these very discreet or atomic presentations that are antinarrative and that resist the pressure on contemporary science, for reasons of funding, to present itself along those narrative lines. You'll find that often when it's a line where the method or the parameters of the experiment are what get discussed but the results or even the hypothesis are left unstated. "Scientists stripped mice of two-thirds of their fur and injected chloroquine in their thoracic spinal cords." That's one where the question would normally be, "What are they trying to test, and what did they actually find out?" But both those things might be more boring than just being reminded of this single event that is equal in value and in form to a goose being shot seven times.

The mouse model is an interesting thing I come back to once in a while. A recent study found that most lab mice are allowed to eat whenever they want, so—among other reasons we already know— it's a really bad comparison for humans. It's like if you were doing a clinical trial on humans who were sitting in a Chipotle twenty-four hours a day and had unlimited credit at the counter. You're bored 'cause you're sitting there in the absence of rich stimuli and all you can do is, say, read your book and keep ordering gigantic burritos. Are those the people on whom you want to be basing important medical research?

It's interesting to see these stupid popular over-generalizations. And also the way that things have to be explained to sell them. Sometimes it's nice to make science opaque and unsexy, whether that's by interrogating the method or by including some really obscure part of what was found. Or by the refusal to explain difficult terms, because there's also this idea that, aside from pure math, anything that anyone does should be explicable, especially to Americans. Clearly that's not true of everything that goes on in physics and chemistry and genetics.

I mean, some of it is just really dense difficult stuff and it's just a pain in the ass to make yourself understand it, and even if you understand it, it may not be very interesting, or it may not have large implications. And sometimes it's fun just to leave those in their radiant opacity.

Where does Findings belong in today's reading landscape?
It can easily be broken up and turned into many iterations and promotions of itself across different social media platforms. It works well on the Web. It works well for short attention spans. A finding is the length of a tweet or a text message. So that's one thing that I think will help keep it in place. You learn things from it that are useful to regurgitate in cocktail-party form. It's excellent toilet reading. Each month may have a different tone, a different set of concerns. You can sometimes tell my emotional state, especially from the last paragraph.

That's one of the things that works for Findings, is that it's so clear in the selection of facts and the wording and the tone that it has a more singular personality or authorship than, say, the Index—which seems group-compiled and voiceless. The fact that it is, like you say, a prose poem, that it's three paragraphs and not a list. Those things alone seem to do so much to add a kind of character to it.

I think an unbylined house voice, or a house voice that could as well be unbylined—it's hard to get to that point. You find it in interesting places. I mean the *Economist*, obviously. But then you have places like the *New Yorker* or Gawker that have a strong enough orientation that in particular kinds of reporting or cultural criticism you could remove the bylines and it would just be a great piece of *New Yorker* prose or a great, incisive, nasty Gawker post. And it's hard for institutions to develop those. It takes time and it takes innovation. You have to be the place that perfected it. At its very best, it inspires imitation, it inspires reevaluation.

Can you talk more specifically about this possible influence of Findings' voice on science writing, or nonfiction in general?

I think Findings is something that in its synthe-sizing capacity, in its collations and its curations and its ironies, is aware of and is sensitive to the dangers of a certain kind of expository and confes-sional sincerity in nonfiction writing generally. It's so intensively and deliberately not about the self. And yet Findings, even as it resists that earnestness and that omphalic gaze, is still a very self-involved thing. It's a self-involved thing that doesn't have a byline. Even as it resists the easy triumphalist narratives and the gee-whiz wonderification of any kind of disposable scientific discovery, it occasionally buys into that completely. It buys into it and celebrates it.

And it's influenced the way I've approached other, longer prose projects—that technique of synthesis and deadpan condensation that's highly editorialized but doesn't seem that way on the surface. But, again, it shows a kind of nonfiction writing that's radically different from this explicit privileging of the self for which I think there's typically too low a bar. I mean nonfiction writing in the therapeutic vein. There are many great practitioners of personal nonfiction and I don't want to dump on anyone's projects, but one thing Findings shows is that you can be deeply self-indulgent without being so deeply *self*. That's something that I would like to see more of.

I guess a lesson that I've learned from writing the column is that you can always say much, much less and still get everything across. Now part of that's a little fallacious or easy for me to say because it's such a condensed form, but the thing I mean more broadly is that rigorous juxtapositions and an economy of language, when you're trying to explain a discovery or a feeling or an irony or a sadness or some sense of joy or absurdity, can convey a lot. If Findings teaches one person to turn in something at 3,000 words and not 4,500 because the main narrative episode is compressed by 50 percent, then that would be a great thing and I would feel that it had a worthy and useful effect on the culture.

Acknowledgments

Vegetarianism does not always cause brain shrinkage.

Citations

Page 1: Morgan K. Ward and Darren W. Dahl. "Should the Devil Sell Prada? Retail rejection increases aspiring consumers' desire for the brand." *Journal of Consumer Research* (June 20, 2014).

Page 2: Lawrence J. Sanna, Edward C. Chang, Paul M. Miceli, Kristjen B. Lundberg. "Rising Up to Higher Virtues: Experiencing elevated physical height uplifts prosocial actions." *Journal of Experimental Social Psychology* (December 22, 2010). (Retraction published January 4, 2013, *Journal of Experimental Social Psychology*.)

Page 3: "Cosmetics, Beauty and Brain Science." Kanebo Cosmetics, press release (October 29, 2009).

Page 4: Birgit Straaso, Lise Lauritzen, Marianne Lunde, Maj Vinberg, Lone Lindberg, Erik Roj Larsen, Steen Dissing, Per Bech. "Dose-Remission of Pulsating Electromagnetic Fields as Augmentation in Therapy-Resistant Depression: A randomized, double-blind controlled study." *Acta Neuropsychiatrica* (April 10, 2014).

Page 5: David G. Moriarty, Matthew M. Zack, James B. Holt, Daniel P. Chapman, Marc A. Safran. "Geographic Patterns of Frequent Mental Distress." *American Journal of Preventive Medicine* (June 2009).

Page 6: Brick Johnstone, Angela Bodling, Dan Cohen, Shawn E. Christ, Andrew Wegrzyn. "Right Parietal Lobe-Related 'Selflessness' as the Neuropsychological Basis of Spiritual Transcendence." *The International Journal for the Psychology of Religion* (September 18, 2012).

Page 7: Björn Vickhoff, Helge Malmgren, Rickard Åström, Gunnar Nyberg, Seth-Reino Ekström, Mathias Engwall, Johan Snygg, Michael Nilsson, Rebecka Jörnsten. "Music Structure Determines Heart Rate Variability of Singers." *Frontiers in Psychology* (July 9, 2013).

Page 8: Vladimir Dinets. "Crane Dances as Play Behaviour." *IBIS: International Journal of Avian Science* (March 16, 2013).

Page 9: Daniel Randles, Steven J. Heine, Nathan Santos. "The Common Pain of Surrealism and Death: Acetaminophen reduces compensation affirmation following meaning threats." *Psychological Science* (April 11, 2013).

Page 10: Igor V. Sokolov, Natalia M. Naumova, John A. Nees, Gérard A. Mouru. "Pair Creation in QED-Strong Pulsed Laser Fields Interacting with Electron Beams." *Physical Review Letters* (November 2010).

Page 11: Olaf Blanke, Polona Pozeg, Masayuki Hara, Lukas Heydrich, Andrea Serino, Akio Yamamoto, Toshiro Higuchi, Roy Salomon, Margitta Seeck, Theodor Landis, Bruno Herbelin, Shahar Arzy, Hannes Bleuler, Giulio Rognini. "Neurological and Robot-Controlled Induction of an Apparition." *Current Biology* (November 6, 2014).

Page 12: Celeste Kidd, Holly Palmeri, Richard N. Aslin. "Rational Snacking: Young children's decision-making on the marshmallow task is moderated by beliefs about environmental reliability." *Cognition* (October 9, 2012).

Page 13: Penny Curtis. "Space to Care: Children's perceptions of spatial aspects of hospitals." ESRC End of Award Report (August 8, 2007).
"Hospital clown images 'too scary.'" BBC News (January 15, 2008).

Page 14: Martie L. Skinner, Kevin P. Haggerty, Charles B. Fleming, Richard F. Catalano. "Predicting Functional Resilience Among Young-Adult Children of Opiate-Dependent Parents." *Journal of Adolescent Health* (October 29, 2008).

Page 15: Amedeo D'Angiulli, Patricia Mana Van Roon, Joanne Weinberg, Tim F. Oberlander, Ruth E. Grunau, Clyde Hertzman, Stefania Maggi. "Frontal EEG/ERP Correlates of Attentional Processes, Cortisol and Motivational States in Adolescents from Lower and Higher Socioeconomic Status." *Frontiers in Human Neuroscience* (November 19, 2012).

Page 17: Mhairi A. Gibson and Rebecca Sear. "Does Wealth Increase Parental Investment Biases in Child Education?" *Current Anthropology* (October 2010).

Page 18: Kristen Copeland, Susan N. Sherman, Cassandra A. Kendeigh, Brian E. Saelens, Heidi J. Kalkwarf. "Flip Flops, Dress Clothes, and No Coat: Clothing barriers to children's physical activity in child-care centers identified from a qualitative study." *International Journal of Behavioral Nutrition and Physical Activity* (November 6, 2009).

Page 19: Samantha J. Sharpe, Lynne M. Rochette, Gary A. Smith. "Pediatric Battery-Related Emergency Department Visits in the United States, 1990–2009." *Pediatrics* (May 14, 2012).

Page 20: Celia M. Lescano, Christopher D. Houck, Larry K. Brown, Glenn Doherty, Ralph J. DiClemente, M. Isabel Fernandez, David Pugarch, William E. Schlenger, Barbara J. Silver. "Correlates of Heterosexual Anal Intercourse Among At-Risk Adolescents and Young Adults." *American Journal of Public Health* (December 2008).

Page 21: Natalie Wolchover. "'Magnetic Boy' Ivan Just a Very Sticky Kid." *LiveScience* (May 11, 2011).

Page 22: Michael L. M. Murphy, George M. Slavich, Nicolas Rohleder, Gregory E. Miller. "Targeted Rejection

Triggers Differential Pro- and Anti-Inflammatory Gene Expression in Adolescents as a Function of Social Status." *Clinical Psychological Science* (September 7, 2012).

Page 23: J. Kiley Hamlin, Karen Wynn, Paul Bloom, Neha Mahajah. "How Infants and Toddlers React to Antisocial Others." *Proceedings of the National Academy of Sciences* (November 28, 2011).

Page 24: Sara J. Solnick and David Hemenway. "The 'Twinkie Defense': The relationship between carbonated non-diet soft drinks and violence perpetration among Boston high school students." *Injury Prevention* (October 24, 2011).

Page 25: Gabriela Guerrero-Serdán. "The Effects of the War in Iraq on Nutrition and Health: An analysis using anthropometric outcomes of children." Households in Conflict Network Working Paper 55 (February 2009).

Page 26: Christoph Bartneck, Karolina Zawieska, Mohammad Obaid. "Agents With Faces—What can we learn from LEGO minifigures?" Proceedings of the 1st International Conference on Human-Agent Interaction (2013).

Page 27: Zinnia J. Janif, Robert C. Brooks, Barnaby J. Dixson. "Negative Frequency-Dependent Preferences and Variation in Male Facial Hair." *Biology Letters* (April 16, 2014).

Page 28: Joerns Fickel, Oleg A. Bubliy, Anja Stache, Tanja Noventa, Adam Jirsa, Marco Heurich. "Crossing the Border? Structure of the red deer population from the Bavarian–Bohemian forest ecosystem." *Mammalian Biology* (November 15, 2011).

Page 30: Susanne Sternthal. "Moscow's stray dogs." *FT Magazine* (January 16, 2010).

Page 31: Nina Mandell. "Earless Bunny Raises Fear of Effects of Nuclear Plant Disaster in Fukushima, Japan." *New York Daily News* (June 9, 2011).

Page 32: Dan R. Klinka and Thomas E. Reimchen. "Adaptive Coat Colour Polymorphism in the Kermode Bear of Coastal British Columbia." *Biological Journal of the Linnean Society* (October 27, 2009).

Page 33: R. J. Nevle, D. K. Bird, W. F. Ruddiman, R. A. Dull. "Neotropical Human–Landscape Interactions, Fire, and Atmospheric CO_2 During European Conquest." *The Holocene* (June 13, 2011).

Page 34: Paul Davies. "Shadow Life: Life as we don't yet know it." Paper presented at the annual meeting of the American Association for the Advancement of Science (February 15, 2009).

Page 35: Zunyi Yang, Fei Chen, J. Brian Alvarado, Steven A. Benner. "Amplification, Mutation, and Sequencing of a Six-Letter Synthetic Genetic System." *Journal of the American Chemical Society* (August 15, 2011).
James Morgan. "Alien Life 'May Exist Among Us.'" BBC News (February 16, 2009).

Page 36: Jennifer S. Mascaro, Patrick D. Hackett, James K. Rilling. "Testicular Volume Is Inversely Correlated with Nurturing-Related Brain Activity in Human Fathers." *Proceedings of the National Academy of Sciences* (September 9, 2013).

Page 37: Kevin M. Williams, Justin H. Park, Martijn B. Wieling. "The Face Reveals Athletic Flair: Better National Football League quarterbacks are better looking." *Personality and Individual Differences* (October 17, 2009).

Page 38: Robert P. Burriss, Hannah M. Rowland, Anthony C. Little. "Facial Scarring Enhances Men's Attractiveness for Short-Term Relationships." *Personality and Individual Differences* (November 8, 2008).

Page 39: Tatiana Levkovich, Theofilos Poutahidis, Christopher Smillie, Bernard J. Varian, Yassin M. Ibrahim, Jessica R. Lakritz, Eric J. Alm, Susan E. Erdman. "Probiotic Bacteria Induce a 'Glow of Health.'" *PLoS ONE* (December 5, 2012).

Page 40: Ken P. Aplin and Muse Opiang. "The Mammal Fauna of the Nakanai Mountains, East New Britain Province, Papua New Guinea." RAP Bulletin of Biological Assessment (May 2009).
"Scientists Discover 200 New Species in Remote PNG." AFP (October 6, 2010).

Page 41: Robert L. Nudds and Gareth J. Dyke. "Forelimb Posture in Dinosaurs and the Evolution of the Avian Flapping Flight-Stroke." *Evolution: International Journal of Organic Evolution* (January 14, 2009).

Page 42: Fabian de Kok-Mercado, Michael Habib, Tim Phelps, Lydia Gregg, Philippe Gailloud. "Adaptations of the Owl's Cervical & Cephalic Arteries in Relation to Extreme Neck Rotation." *Science* (February 1, 2013).

"Owl Mystery Unraveled: Scientists Explain How Bird Can Rotate Its Head Without Cutting Off Blood Supply to Brain." Johns Hopkins Medicine, press release (January 31, 2013).

Page 43: "Plum Pox Virus, Stone Fruit—Japan: Spread." ProMED-mail (June 29, 2012).

"Avian Pox—USA (03): (Arizona) Hummingbird, Suspected." ProMED-mail (July 2, 2012).

Page 44: Michael Fröhlingsdorf. "Bird-Brained Idea: Vulture detective training hits headwinds." *Spiegel Online International* (June 28, 2011).

Page 45: Andrew D. Wade, Salima Ikram, Gerald Conlogue, Ronald Beckett, Andrew J. Nelson, Roger Colten, Barbara Lawson, Donatella Tampieri. "Foodstuff Placement in Ibis Mummies and the Role of Viscera in Embalming." *Journal of Archaeological Science* (January 13, 2012).

Page 46: "'Oldest sheep' in the World Dies on Lewis in Western Isles." BBC News (February 28, 2012).

Page 47: Adam Ostrzenski. "G-Spot Anatomy: A new discovery." *The Journal of Sexual Medicine* (April 25, 2012).

Page 48: Anthony J. Sanford, Hartmut Leuthold, Jason Bohan, Alison J. S. Sanford. "Anomalies at the Borderline of Awareness: An ERP study." *Journal of Cognitive Neuroscience* (December 9, 2010).

Page 49: "Giving Old Eyes a Tune-Up." The Vision Centre, press release (April 2009).

Page 50: "Almost One Third of London Men Are Too Fat to See Their Own Genitals, Study Finds." *London Evening Standard* (October 1, 2012).

Page 51: Javaad Zargooshi. "Sexual Function and Tunica Albuginea Wound Healing Following Penile Fracture: An 18-year follow-up study of 352 patients from Kermanshah, Iran." *The Journal of Sexual Medicine* (December 4, 2008).
Regina Nuzzo. "Preventing Penile Fractures and Peyronie's Disease." *Los Angeles Times* (February 9, 2009).

Page 52: Keith Jeffery. *MI6: The History of the Secret Intelligence Service, 1909–1948*. London: Bloomsbury, 2010.

Page 53: Romana Khan, Kanishka Misra, Vishal Singh. "Ideology and Brand Consumption." *Psychological Science* (February 4, 2013).

Page 54: B. Wansink and C. S. Wansink. "The Largest Last Supper: Depictions of food portions and plate size increased over the millennium." *International Journal of Obesity* (March 23, 2010).

Page 55: Roelof van der Broek. *Pseudo-Cyril of Jerusalem on the Life and the Passion of Christ: A Coptic Apocryphon.* Boston: Brill, 2012.

Page 56: D. Roberts, J. Nelms, D. Starkey, S. Thomas. "Travelling by Teleportation." *Journal of Physics Special Topics* (November 6, 2012).

Page 57: Lenka Halámková, Jan Halámek, Vera Bocharova, Alon Szczupak, Lital Alfonta, Evgeny Katz. "Implanted Biofuel Cell Operating in a Living Snail." *Journal of the American Chemical Society* (March 8, 2012).

Page 58: "EATR: Energetically Autonomous Tactical Robot." Robotic Technology, Inc., press release (April 20, 2009).
Bobbie Johnson. "'Flesh-Eating Robot' Is Actually a Vegetarian, Say Inventors." *The Guardian* (July 19, 2009).

Page 59: "Worm-Eating Slug Found in Garden." BBC News (July 10, 2008).

Page 60: E. Sunamura, X. Espadaler, H. Sakamoto, S. Suzuki, M. Terayama, S. Tatsuki. "Intercontinental Union of Argentine Ants: Behavioral relationships among introduced populations in Europe, North America, and Asia." *Insectes Sociaux* (March 4, 2009).

Page 61: Ridwanur Rahman, M. Abul Faiz, Shahjada Selim, Bayzidur Rahman, Ariful Basher, Alison Jones, Catherine d'Este, Moazzem Hossain, Ziaul Islam, Habib Ahmed, Abul Hasnat Milton. "Annual Incidence of Snake Bite in Rural Bangladesh." *PLoS Neglected Tropical Diseases* (October 26, 2010).

Page 62: Ruben Andres Berthé, Stéphanie de Pury, Horst Bleckmann, Guido Westhoff. "Spitting Cobras Adjust Their Venom Distribution to Target Distance." *Journal of Comparative Physiology* (May 22, 2009).

Page 63: Amanda M. Vicary and R. Chris Fraley. "Captured by True Crime: Why are women drawn to tales of rape, murder, and serial killers?" *Social Psychological and Personality Science* (January 2010).

Page 64: Scott R. Loarie, Craig J. Tambling, Gregory P. Asner. "Lion Hunting Behaviour and Vegetation Structure in an African Savanna." *Animal Behaviour* (May 2013).

Page 66: Arvid Guterstam, Valeria I. Petkova, H. Henrik Ehrsson. "The Illusion of Owning a Third Arm." *PLoS ONE* (February 23, 2011).

Page 67: Colin D. Kennedy, Stephen W. Houmes, Katherine L. Wyrick, Samuel M. Kammerzell, Ken Lukowiak, Barbara A. Sorg. "Methamphetamine Enhances Memory of Operantly Conditioned Respiratory Behavior in the Snail *Lymnaea stagnalis.*" *The Journal of Experimental Biology* (March 3, 2010).

Page 68: Piotr G. Jablonski, Hyun Jun Cho, Soo Rim Song, Chang Ku Kang, Sang-im Lee. "Warning Signals Confer Advantage to Prey in Competition with Predators: Bumblebees steal nests from insectivorous birds." *Behavioral Ecology and Sociobiology* (May 24, 2013).

Page 70: "Scaredy-Cat Tigers." Ananova (November 20, 2009).

Page 71: A. M. Overduin-de Vries, J.J.M. Massen, B. M. Spruijt, E.H.M. Sterck. "Sneaky Monkeys: An audience effect of male rhesus macaques (*Macaca mulatta*) on Sexual Behavior." *American Journal of Primatology* (January 23, 2012).

Page 72: Kirsten M. Bohn, Barbara Schmidt-French, Christine Schwartz, Michael Smotherman, George D. Pollak. "Versatility and Stereotypy of Free-Tailed Bat Songs." *PLoS ONE* (August 25, 2009).

Page 73: A. Avargues-Weber, G. Portelli, J. Bernard, A. Dyer, M. Giurfa. "Configural Processing Enables Discrimination and Categorization of Face-Like Stimuli in Honeybees." *The Journal of Experimental Biology* (November 16, 2009).

Page 74: Helene Hopfer, Susan E. Ebeler, Hildegarde Heymann. "The Combined Effects of Storage Temperature and Packaging Type on the Sensory and Chemical Properties of Chardonnay." *Journal of Agricultural and Food Chemistry* (October 4, 2012).

About the Author

RAFIL KROLL-ZAIDI

Rafil Kroll-Zaidi is a contributing editor of *Harper's Magazine* and has been writing its popular Findings column, among other features, since 2007.

About the Artist

GRAHAM ROUMIEU

Graham Roumieu is the author and illustrator of the celebrated Bigfoot "autobiographies" *In Me Own Words, Me Write Book,* and *I Not Dead.* His drawings appear in such publications as the *Atlantic, Harper's Magazine,* the *New York Times,* and the *Walrus.*

About the Foreworder

PATTON OSWALT

Patton Oswalt is a comedian when he isn't writing books or acting in things or drinking coffee.

ABOUT TWELVE

TWELVE

TWELVE was established in August 2005 with the objective of publishing no more than twelve books each year. We strive to publish the singular book, by authors who have a unique perspective and compelling authority. Works that explain our culture; that illuminate, inspire, provoke, and entertain. We seek to establish communities of conversation surrounding our books. Talented authors deserve attention not only from publishers, but from readers as well. To sell the book is only the beginning of our mission. To build avid audiences of readers who are enriched by these works—that is our ultimate purpose.

For more information about forthcoming TWELVE books, please go to www.twelvebooks.com.